YOUR KNOWLEDGE HAS VALUE

- We will publish your bachelor's and
  master's thesis, essays and papers

- Your own eBook and book -
  sold worldwide in all relevant shops

- Earn money with each sale

Upload your text at www.GRIN.com
and publish for free

Markus Biedermann

Aviation Biofuels. Curse or Cure for Green Aviation?

GRIN Publishing

Bibliographic information published by the German National Library:

The German National Library lists this publication in the National Bibliography; detailed bibliographic data are available on the Internet at http://dnb.dnb.de .

Imprint:

Copyright © 2015 GRIN Verlag GmbH
Print and binding: Books on Demand GmbH, Norderstedt Germany
ISBN: 978-3-656-92508-8

This book at GRIN:

http://www.grin.com/en/e-book/294668/aviation-biofuels-curse-or-cure-for-green-aviation

MBA Aviation Management

Module: Economic and Environmental Basis, Applied Regional Analysis
and Spatial Economics

SS 2014

Aviation Biofuels

Curse or cure for green aviation?

by

Markus Biedermann

[November 24th, 2014]

Inhaltsverzeichnis

List of Figures

Abbreviations

BtL	Biomass-to-Liquid
CO2	carbon dioxide
EU	European Union
EU ETS	European Union Emissions Trading Scheme
FT	Fischer-Tropsch
HEFA	Hydro-processed Esters and Fatty Acids
Mboe/d	Million barrel of oil equivalent per day
TTW	Tank-to-Wheel
WTT	Well-to-Tank
WTW	Well-to-Wheel
XtL	Factor X-to-Liquid

Abstract

The purpose of this paper is to evaluate the challenges and chances of biofuels in an aviation context. It begins by providing a brief history on biofuels in general and in aviation specifically. It is followed by an overview of the main driving forces for the aviation industry to invest in biofuels and their development. After that, different available biofuels and their segmentation in first, second, and third generations is shown in more detail along with a basic overview on the production processes that exist on the market. Based on the given information, first the challenges and after that, the chances, for the aviation industry on investing in aviation biofuels is shown into detail.

The paper claims that the biofuels still have major issues as there is no real alternative on the market or in development that can substitute crude oil-based jet fuel up to 100%. Current biofuels can only be used as mixtures with conventional jet fuel. However, the biggest issue, namely that biofuels cannot be certified in aviation because of their specifications, is solved with second-generation biofuels.

The main challenges are still to find a sustainable biofuel that can be used without changes to existing aircraft propulsion technology and nothing ensures that this will be possible in the necessary timeframe until the crude oil finally starts to become less available. That point in time is constantly updated, but only based on statistical data. It is not known at what point in time this will be exactly.

However, there are a number of opportunities. They range from having an alternative to substitute conventional jet fuel for emission reduction reasons to the investment of an aircraft operator to the production of biofuel itself. This could be for the sake of entering new markets or just to integrate the fuel supply into the company-owned supply chain. The paper concludes that a lot of development still seems to need to be required before a real alternative to conventional jet fuel is available on a large scale. The most promising approach today is the algae-based fuel

production, which uses genetically modified algae to directly produce an
oil derivative.

1. Introduction

Fuel made from crude oil has been vital for the aviation industry since the beginning of commercial aviation in 1908, when Wilbur Wright took an employee with him in his plane. (BURNS&MCDONNELL, 2014) Today, the world oil demand in aviation has reached 5.5 million barrels of oil equivalent per day (mboe/d) and due to the growth in the aviation industry, this demand is still rising with a projected demand of 7.2 mboe/d by 2035. (ORGANIZATION OF THE PETROLEUM EXPORTING COUNTRIES, 2013) Jet fuel price right now is decreasing and reached a minus of 16.5% in average if the price level of October 2013 and October 2014 is compared. (INTERNATIONAL AVIATION TRANSPORT ASSOCIATION, 2014) However, the overall fuel price in the past few years has risen remarkably from 104.1 USD per barrel in December of 2010 to 121.5 USD per barrel in July of 2014. That is an overall increase of approximately 16.7% in just four years. (INTERNATIONAL AVIATION TRANSPORT ASSOCIATION, 2014) For that reason alone, the search for alternative aviation fuels is interesting, since from an economical point of view, there are alternatives to cut costs. On the other hand, the continuing increase in the awareness about carbon dioxide and its implications on the climate is putting additional pressure on the search for alternatives.

This paper focuses on biofuels in aviation and what possible chances and challenges it has in the aviation industry as an alternative to crude oil-based fuels. It starts by providing an overview of the development of biofuels in general and for aviation specifically. Based on that, the associated challenges are evaluated and thus, so too are the possible chances that result for the aviation industry. Upon conclusion, an outlook is provided for possible further developments and options in the near future.

2. Definition of Biofuel

Biofuels are a portion of bioenergy or biomass energy that is produced from various organic materials. The production relies on biological processes and this basically means that any available organic material is used to produce liquid fuel that is then known as ethanol, methanol, and biodiesel. Especially important is that biofuels contain energy from previous carbon fixations, which means that burning that sort of fuel only sets free the amount of carbon dioxides that have been fixated during the biomass formation process. (OLAH, George A. et al., 2009)

3. History of Biofuels

Until the industrial revolution in the 18th century, the predominant source of energy had been biomass materials such as wood, plant oil, or animal fat. During the industrial revolution, the former bioenergy sources were replaced with fossil fuels, coal, and later oil and gas in a step by step process. (BORCHARDT, Knut, 1978) In the early 1980s, new research on renewable energy sources started after the initial oil crises in the 1970s. (KARLSCH, Rainer and Stokes, Raymond G., 2003) The world became aware of the fact that crude oil, as well as coal and gas, are finite as the basis of energy production. At some unknown point in the future, and even in the current day and age, the calculations will differ remarkably and fossil sources for energy will be completely consumed.

The developments at that time allowed for the first generation biofuels that have since been used primarily heating homes and power generation. The most common source for the production of those biofuels is edible plants that are rich in sugars and bio-derived oils. The sugar-containing plants can be fermented to produce ethanol as a direct substitute for petroleum.

The second generation biofuels are based on bio-derived oil that is usually produced through pressing the plants to get the oil. Bio-derived oil can be burnt directly or processed further to make jet and diesel fuels. The second-generation biofuels were introduced during the 1990s. (AIR TRANSPORT ACTION GROUP, 2009) Most of those biofuels are still produced from different feedstock.

The latest development in biofuels and especially in bio jet fuels are the third generation biofuels that rely on certain plants like algae that are genetically manipulated to produce different types of fuel. The research is still in its initial stages and no long-term experience exists at this point as only research farms are currently in existence. Availability for commercial use is to be expected as of 2035 and onwards. (BIOFUEL.ORG.UK, 2014)

4. Driving forces for aviation biofuels

4.1. Economical reasons

Research on biofuels for aviation already started with first generation biofuels. At the beginning, the interest in the industry was mainly derived from the perspective to gain independence from oil producing countries and then increase the bargaining power. With the oil crisis, the volatility of aviation fuel prices has been an important factor in establishing the first approaches into aviation biofuels. For the aviation industry, fuel costs are one of the main expenses of operations. In 2013, Lufthansa spent 22.5% of the overall operating expenses on fuel, which in total numbers is 7.058 million EUR. (LUFTHANSA, 2013) Even just little changes in the overall aviation fuel price can contribute greatly to the operating expenses. An alternative source for aviation fuel that is not dependent on the oil price would remove the pressure on the fuel policies of airlines.

4.2. European Union Regulation

The second main driving force is the green house discussion that started in the early 1990s. The natural green house effect is comprised of the release of carbon in the form of carbon dioxide into the atmosphere and then becomes fixated in soil via plants that absorb this carbon dioxide. This is called the natural carbon cycle.

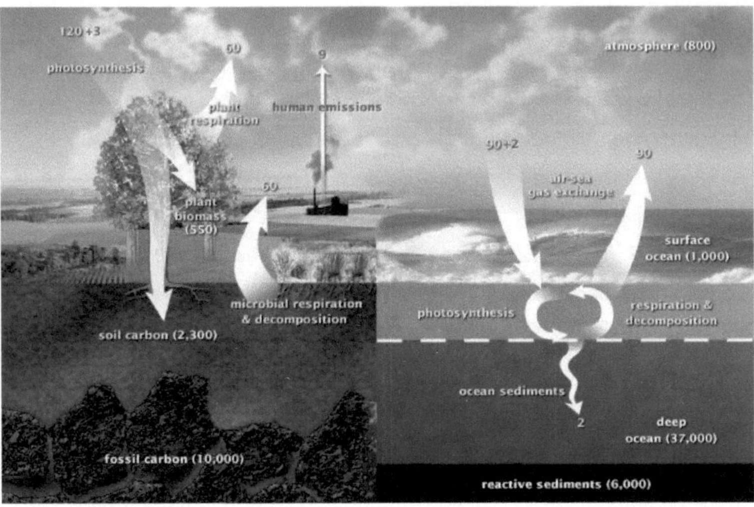

Figure 1: The Carbon Cycle (2014)

Through the burning of fossil fuels, additional carbon is released into the atmosphere and cannot be fixated by the existing biomass. That effect increases the greenhouse effect, which is another term for global warming. (KUMP, Lee R. et al., 2009)

Although aviation only contributes roughly 2% of the world's carbon dioxide emission (LEE, D.S. et al., 2010) and 12% of the carbon dioxide emissions in the global transport sector, the public often perceives it as being much higher than that, because ultimately, aviation is a huge worldwide business that is highly visible in the media.

This has lead to the European Union Emissions Trading System (EU ETS), which is a regulatory tool that was implemented in 2005 with the goal to reduce the total emission of greenhouse gases. Even though this only applies to the 28 EU member states as well as Iceland, Liechtenstein, and Norway, it has a global impact. The EU ETS not only applies to operators located within those countries, but also to any operator of flights to and from EU ETS participating countries. (EUROPEAN COMMISSION, 2013)

The system's goal is to reduce greenhouse gas emissions by 20% until 2020 and by 80-95% by 2050 in comparison with the levels as they were in 1990. However, there is no definition on to go about reducing these emissions. (EUROPEAN COMMISSION, 2013)

The EU ETS is divided in three phases or trading periods. The first phase is derived from the Kyoto Protocol development, which started in 1998, but has already been in place since before the protocol came into force in February of 2005. (UNITED NATIONS, 1998) The first phase was in place from 2005 until 2007 and functioned as a pilot period. In that phase, aviation was not part of the trading scheme as only power generators and energy-intensive industrial sectors were affected. The allowances have been given free of charge to the businesses and the penalty for non-compliance has been set to 40€ per ton of carbon dioxide (CO_2). However, none of Iceland, Liechtenstein or Norway participated in phase one. (EUROPEAN COMISSION, 2014)

The second period, which lasted from 2008 until 2012 included not only CO_2 emissions, but also nitrous oxide emissions by a number of member states and the penalty for non-compliance increased to 100€ per ton. The first auctions for the allowances were held during phase two. (EUROPEAN COMISSION, 2014) Airlines were able to trade missing allowances from airlines that did not need everything they had. Within this period, the 2008 economic crisis took place and many allowances have gone unused.

Phase three then actually started in 2013. The major change from the previous phases was that the allowances were no longer issued on a national basis, but rather the EU commission issued an EU-wide upper limit to the overall CO_2 emissions that have to be met jointly within the member states. In addition, the initial issue of allowances was no longer issued for free, but also on the basis of auctions. (EUROPEAN COMMISSION, 2013)

Especially aviation has been taken into consideration under the guise of EU ETS from 2012. Thus, this has been at the end of the second phase and it did not affect the second phase, but did have an impact on the entire third phase completely. In addition, operators outside the EU such as Russian and Chinese operators are fighting against European regulation authorities.

5. Available aviation biofuel approaches

The current research on alternative jet fuels and especially biofuels for aviation is developing in six different approaches that result in nine different usable biofuel alternatives for the aviation industry. The first generation biofuels are derived from gas, coal, or sugar-based crops. This is achieved by using the Fischer-Tropsch (FT) process. Second generation biofuels are produced via the Biomass-to-Liquid (BTL) process from urban waste and Lignocelulose, such as wood and agricultural products. The third generation aviation biofuels are generated with the hydro-processed esters and fatty acids (HEFA). (KÖHLER, Jonathan et al., 2013) Until now, the only drop-in option consists of hydrocarbons, since only those products can be processed to function in current jet engines.

5.1. Fischer-Tropsch and the Biomass-to-Liquid process

The FT process developed by Franz Fischer and Hans Tropsch in 1925 makes use of coal and natural gas to produce hydrocarbons that can be further processed to aviation fuels. The usage of this process and the jet fuels derived by it was approved for aviation use in 2009. (INTERNATIONAL AIR TRANSPORT ASSOCIATION, 2013)

In addition, the FT process also serves as the basis of the BTL process. This can turn biomass coming from whole plants or urban waste into hydrocarbons and then into aviation biofuels from there. The products are also named XtL fuels, where 'X' stands for a variable and 'tL' is for "to Liquid". The acronym is the short version for the FT process that goes about turning any X into a gas form and further chemical processing into liquid fuel forms. If biomass is turned into fuel, it would the already mentioned BtL process. For the BTL process used to gain biofuel for aviation use, one especially has to use switch grass, corn stover, and forest residues. Urban waste would be feasible as well, but has proven to be more expensive to establish than using plants.

The limits of this process are however, that it is not possible to generate a product that can supplement classic jet fuel 100%. It always has to be a blend of things. (KÖHLER, Jonathan et al., 2013)

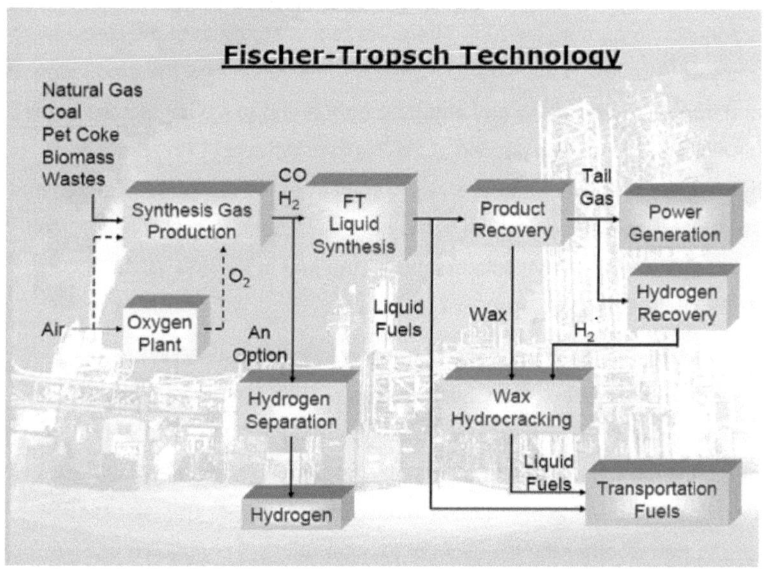

Figure 2: Fischer-Tropsch Technology (2014)

5.2. Hydro-processed esters and fatty acids

HEFA jet fuels are also referred to as renewable oil processing. In this approach, renewable oils such as those from algae are processed into a fuel that has similar properties to those in the Fischer-Tropsch process, but they can also be used without having to be mixed in a direct drop-in. Apart from algae, it's also feasible to make use of soybeans, palm oil, jatropha, salicornia, and rapeseed for fuel production. Tests with mixtures of 50% HEFA jet fuel and 50% conventional fuels have already been successfully finished. Especially in terms of CO2 reduction over the entire lifecycle of such a jet fuel, the algae-based HEFA jet fuel is most promising.

5.3. The carbon life cycle

In terms of green aviation and the CO2 discussion, we need to take more than just CO2 emissions caused by burning fuel into account. We also need to look at the whole lifecycle from gathering the basis for fuel to production to burning the fuel. There are two major parts of the lifecycle analysis. The first is the Well-to-Tank (WTT) emissions of the production and transportation of the fuel and then right on up to it being placed in the tank of an aircraft. The second is the Tank-to-Wheels (TTW) process of burning the fuel. Since the analysis comes from the car industry, acronyms are used in reference to wheels and not to jet engines. The analysis itself is independent of the vehicle that burns the fuel at the end. (JOINT RESEARCH CENTRE, 2014)

The lifecycle assessment based on the Well-to-Wheel (WTW) method takes into account just how much carbon is stored in the basis of future fuel and how much carbon is emitted in terms of CO2 during the process of turning the basis into fuel and finally the burning of the fuel itself. As a baseline, the burning of crude oil is used in this model. The assessment shows that whether the life cycle of an alternative to crude oil emits – in total – more or less CO2 than crude oil. (JOINT RESEARCH CENTRE, 2014)

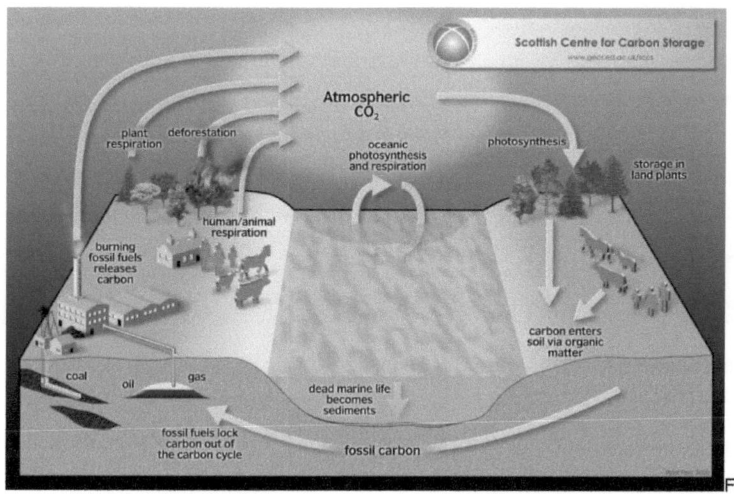

Figure 3: The Carbon Life Cycle (2014)

6. Challenges for aviation biofuels

Biofuels in aviation are one major way to tackle the challenges of the future in the worldwide growing aviation market. However, biofuels are not free of problems. The challenges hold true for every generation of biofuels, however the newer the generation of technology is, the better the challenges are dealt with.

6.1. Food production competition

Sugar crops, soybeans, palm trees, and rapeseed are, for example, all edible plants. If being used for the production of biofuels, they are no longer available for feeding. In addition, huge farmlands are necessary to grow enough plants for the demands of the fuel producing industry. The huge demand in fertile land puts biofuel production into direct competition to the agricultural industry. On the one hand, the plants themselves could be used to feed people and serve the basic demands of humanity. If an edible plant is grown to produce biofuels, it directly competes with the demand on food. On the other hand, only farmland itself is a competition to food production. Farmland that is used for growing biofuel plants is then no longer available for growing other necessary crops. Especially in underdeveloped countries, this can be a big issue since the production of biofuels could be more interesting on an economical scale and the industry sector could change from producing basic demand goods to biofuels, which puts even more pressure on the respective country. (INTERNATIONAL AIR TRANSPORT ASSOCIATION, 2013)

6.2. Land use competition

The competition between farmland for producing biofuels and farmland for growing food is not the only form of competition with biofuels with respect to land use. Since the amount of land needed for producing biofuels is very high, it also competes with land use in general. If a European country would start to produce biofuels on a large scale, the availability of land for housing would be reduced dramatically. (BIOFUEL.ORG.UK, 2014) In many African countries, where even the growing food is minimally possible in certain places, there are often also problems with space for housing, which makes the problem even greater than expected. (MANTZOS, L. and Capros, P., 2006)

6.3. Availability

The availability of biofuel for aviation as a global business is vital if the industry is to start relying on it. Since crude oil can be only be transported through a few places throughout the world, the countries that control the crude oil production will ultimately control the world industry to a certain extent, as a lot of industries rely on the availability of crude oil. (ORGANIZATION OF THE PETROLEUM EXPORTING COUNTRIES, 2013)

If the world starts producing biofuels from plants to a greater extent, it can be done all over the world, which can be counted as an advantage. The difficulty here is that especially countries that most need the produced biofuels are not always able to provide enough land to grow the plants due to their dense population. (KÖHLER, Jonathan et al., 2013) This applies to the aviation industry as well. Still, availability not only relates to the places where it can be produced, but also to how reliable the availability is. If biofuel production relies on plants, the growth of the plants is a vital aspect. With bad weather or bad harvests, the availability could be limited. This would also relate to the price volatility of biofuels.

6.4. Price volatility

One of the economic drivers is the price volatility of crude oil. This is also a potential problem for biofuels. Since the challenge of food production and land use competition might force the biofuel industry to concentrate at some places around the world, such as oil producing countries, this leads to a similar industrial structure with similar effects on the entire industry. Without a widespread availability, it is highly likely that comparable effects happen just as they currently do in the oil producing industry. (AGARWAL, Ramesh K., 2012)

Price volatility is also likely to be influenced by severe weather effects that are more likely to happen with changing climate. Since oil fields are located below the earth's surface, they are not affected by severe weather. The supply of plants for fuel production can be severely endangered by weather effects, which would have a direct impact on both the supply and price. (KARLSCH, Rainer and Stokes, Raymond G., 2003)

6.5. Compatibility and certification

Aviation Biofuels are only feasible if they can be used as drop-in products. That means that the technology of the aircraft does not have to be changed. Existing propulsion technology has to be able to burn the biofuels that are provided ideally to 100% and not only as mixtures with conventional biofuels. In addition, the chemical and physical characteristics of aviation biofuels need to be at least as reliable as conventional aviation fuels. This leads to a certain certification process for aviation biofuels. This process exists so as to secure the safe operation of aircraft. From the many varieties of alternative fuels and biofuels, only a few are able to meet the certification requirements. At the end of the day, this has a direct influence on the availability. (EXXON MOBIL AVIATION, 2005)

6.6. Potential risk of a dead end street

The most imminent problem with drop-in biofuels is that no research that has been conducted takes into account the possibility that it might not be possible to produce a drop-in fuel alternative on a large scale. The research itself is promising, especially with the algae approach, but it still has not proven to be feasible in any manner whatsoever. Manufacturers such as Airbus, which makes use of less fuel consuming aircraft vehicles such as the A320 NEO series, have some alternative products in testing, for example, electrically driven aircraft, but only in at a very early stage in development. The main effort still is only placed on the development of aircraft using conventional engines. (AIRBUS - LEADING AIRCRAFT MANUFACTURER, 2014)

7. Possibilities for aviation biofuels

Despite the various challenges, the aviation industry is investing in aviation biofuels. This is due to the fact that alternative fuels can have huge advantages. This is especially the case in light of the fact that EU ETS are a means in which to solve future challenges in advance.

7.1. CO2 lifecycle and EU ETS

With the EU ETS, the emissions trading scheme has been implemented and the allowances are continuously cut back. At the same time, due to simple economic reasons of supply and demand, the price during the auctions of the allowances then rises. Although this effect is not that fierce at the moment, because the aviation industry is recovering from an economic crisis, it will be a challenge in the near future.

(INTERNATIONAL AIR TRANSPORT ASSOCIATION, 2013) Biofuels with a lifecycle that possibly fixes more carbon than it emits in total would have a positive effect on the necessity of allowances. HEFA-produced aviation biofuels from algae have the potential for that. Genetically modified algae that produce hydrocarbons that can be processed into fuel with little effort help to solve the problems with the EU ETS on a sustainable level. (MANTZOS, L. and Capros, P., 2006)

7.2. Cost effectiveness

With a sustainable biologically produced alternative to conventional jet fuel, the operator of aircraft becomes more and more independent on the price volatility of crude oil. Even with low price levels of jet fuel at the moment compared to a three-year average (INTERNATIONAL AVIATION TRANSPORT ASSOCIATION, 2014), the overall fuel price continuously rises with very strong deviations from the mean. With alternatives, those effects can be covered up to a level where they are essentially cancelled out. This desirable level of effectiveness impacts the cost effectiveness of the airline to a high extent as the hedging policies of airlines can revert to alternative methods from those currently being used. (LUFTHANSA, 2013)

In addition, biofuels have the potential to be cost effective in production and distribution, because e.g. of lower security standards for transport and storage. This would result in an alternative for conventional jet fuel on a lower average price level. (MANTZOS, L. and Capros, P., 2006)

7.3. Independence of fuel producing countries

Another important point to mention is the possible independence of the fuel producing countries. A few countries with possibly unstable political environments are the main source for crude oil worldwide. The possibility of crises and problems is high as is the bargaining power of the producing countries. Until today, there is no large-scale alternative to crude oil and even without the risk of a shortage on fuel in the near future, the risk of a sudden increase in the price level is high. (BIOFUEL.ORG.UK, 2014)

7.4. Entering new markets

While it is not interesting for an airline to invest in the flowing of crude oil or the infrastructure to produce oil because of the related costs and the saturated markets, it can be interesting to invest in biofuels. As just like crude oil, the production of biofuels not only serves the demands of the aviation industry, the ownership or partial ownership of the production, could be used as a form of backward integration. An airline can add the fuel supply to its company structure in order to get better control of its own supply chain. Since the market is not yet settled, the chances of gaining profitable price-to-earnings ratios are high. On the other hand, the integration of technology and companies with a green image amongst the customers in addition to a high level of media visibility adds value to the image of the individual company as well. Both can be used as competitive advantages from an economical point of view.

7.5. Dealing with the crude oil shortage in the future

Crude oil will not be available forever, as it only exists in a specific non-recoverable part of the earth. In the 1960s, M. King Hubbert presented studies on so-called peak oil. This is the point somewhere in the future when the world oil production will reach its peak. From then on, it will constantly decline until there is no more crude oil on earth. (HUBBERT, M. King, 1962) Throughout the years, Hubbert has constantly updated his predictions. Meanwhile, it's not only the peak production that is the focus, but also the point where the world's demand surpasses the world's production. While predictions always have a certain degree of uncertainty because of the method, the basic idea behind the Hubbert prediction is valid. There will be one point somewhere in the future that is not currently well-defined, when oil will be a limiting factor. At that point in time, the industry has to have alternatives at hand in order to totally substitute crude oil. The years leading up to that point will be accompanied by massive price increases on crude oil due to shortage in the supply. The alternatives from renewable sources have to be at hand on a global scale already at that stage so as to oppose that problem. (LEE, D.S. et al., 2010)

7.6. Algae-based jet fuel

The most often mentioned pathway towards achieving a sustainable source of biofuel that does not compete with what is necessary for fostering food or land is algae oil. The algae can be grown in ponds, bioreactors or a combination of both. Since algae are non-edible, they do not compete with the food production process. Since they can be grown on a large scale in water, they do not compete with land use. Especially the growing of algae in ponds is promising for the industry, as it does not entail high production costs. Bioreactors have a higher rate of efficiency, but still at much higher costs. Yet the technology is still under development and further improvement will decrease the production costs significantly. A special type of genetically modified algae is able to produce oil directly. It can be further processed to fuel. This method is new on the market and there is very little known about it thus far. (OLAH, George A. et al., 2009)

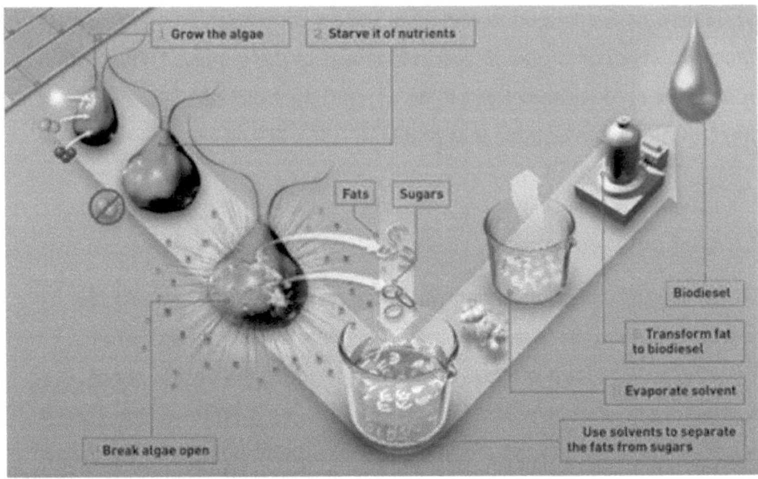

Figure 4: Algae Fuel production (2014)

8. Conclusion

It is without question that the future of aviation needs alternatives to current crude oil-based jet fuels. The availability of biofuels seems to be the right way as it can solve the fuel problem and the climate problem hand in hand. The crude oil will, at some point in time, no longer be available and an industry that is relying on propulsion engines that need to consume energy, this would be a terminal situation. The first generation biofuels are not at all an alternative as they compete with the food and land supply of human beings and also do not fit the needs of the aviation industry. On the other hand, the climate change becomes more and more visible to the public in the form of severe weather phenomena, which also directly influence the whole aviation industry. Operating in severe weather means there's a risk for the safety of the passengers as well as increased fuel consumption as severe weather causes delays, rerouting and inefficient flight paths. Aviation, since it is very close the physical sky, is seen by the public as a major climate killer. Research done about CO2 emissions proves that aviation only contributes on a very little scale to

CO2 emissions. This is not largely recognized and this fact puts pressure on the whole sector.

Looking at the topic in more detail, the efforts made to this point in time are all based on the very old Fischer-Tropsch process that was originally developed to convert coal into liquid fuel. Taking into account that this process is still used on a large scale today just with different input sources, it seems like the research did not yet develop a real alternative. The use of genetically modified algae is the first real new approach on the topic of biofuels. A plant that is capable of producing oil directly without needing the Fischer-Tropsch process sounds promising. On the other hand, the research is still very much in its initial stages and the potential of the genetically modified algae is still a theory that needs to prove its feasibility.

The fact that drop-in biofuels might not be available on a large scale at any point in the future is not taken into account. Manufacturers of aircrafts have done some research on alternative propulsion technology, but only to a very small extent. Reduction of fuel consumption or higher compatibility to biofuels is less costly and more promising from a manufacturer's viewpoint. As such, the imminent risk of failing is present. At this point, the whole industry seems to be narrow-minded. Parallel to those activities, the EU is using the emissions trading scheme to put a lot of pressure on the industry to improve the CO2 efficiency, but the pressure mainly affects the EU operating airlines as there is no worldwide consensus and the competitiveness of the EU airline operators is affected to a certain extent.

With the current development of air traffic, the climate goals are not that difficult to fulfil, but they will get hard to meet in the near future as air traffic is recovering and the climate goals are becoming harder each year. If the industry by itself is not able to give an answer to this question, specific CO2 goals for engines or aircraft are very likely and they would be even more challenging as they could force operators to change whole fleets.

Finally, it is obvious that a lot of effort on different topics is done, but the real game-changing development up to now is neither on the rise nor already on the table, even though some promising approaches are available. The next ten years most likely will show the direction that biofuels will take in the future. From the perspective of this paper, there are only two options left. Either a game-changing drop-in fuel will be presented or an alternative way to move an aircraft will have to be developed.

Bibliography

AGARWAL, Ramesh K. 2012. *Review of Technologies to Achieve Sustainable (Green) Aviation.* St. Louis.

AIR TRANSPORT ACTION GROUP. 2009. *Beginner's Guide to Aviation Biofuels.* Geneva.

AIRBUS - LEADING AIRCRAFT MANUFACTURER. 2014. *Eco-efficiency at every step of the aircraft life-cycle.* [online]. [Accessed 21 Oct 2014]. Available from World Wide Web: < HYPERLINK "http://www.airbus.com/innovation/eco-efficiency/design/fuel-cells/" http://www.airbus.com/innovation/eco-efficiency/design/fuel-cells/ >

BIOFUEL.ORG.UK. 2014. *Biofuels - The fuel of the future.* [online]. [Accessed 16 Oct 2014]. Available from World Wide Web: < HYPERLINK "http://biofuel.org.uk/third-generation-biofuels.html" http://biofuel.org.uk/third-generation-biofuels.html >

BORCHARDT, Knut. 1978. *Grundriß der deutschen Wirtschaftsgeschichte.* Göttingen: Vandenhoeck & Ruprecht.

BURNS&MCDONNELL. 2014. *Timeline of Commercial Aviation.* [online]. [Accessed 16 Oct 2014]. Available from World Wide Web: < HYPERLINK "http://www.burnsmcd.com/Aviation-Special-Report/Article/Timeline-of-Commercial-Aviation" http://www.burnsmcd.com/Aviation-Special-Report/Article/Timeline-of-Commercial-Aviation >

EUROPEAN COMISSION. 2014. *European Comission Climate Action.* [online]. [Accessed 17 Oct 2014]. Available from World Wide Web: < HYPERLINK "http://ec.europa.eu/clima/policies/ets/pre2013/index_en.htm" http://ec.europa.eu/clima/policies/ets/pre2013/index_en.htm >

EUROPEAN COMISSION. 2014. *European Comission Climate Action.* [online]. [Accessed 17 Oct 2014]. Available from World Wide Web: < HYPERLINK

"http://ec.europa.eu/clima/policies/ets/pre2013/second/index_en.htm"
http://ec.europa.eu/clima/policies/ets/pre2013/second/index_en.htm >

EUROPEAN COMMISSION. 2013. *The EU Emissions Trading System (EU ETS)*. Brussels.

EXXON MOBIL AVIATION. 2005. *World Jet Fuel Specifications*.

HUBBERT, M. King. 1962. *Energy Resources*. New York.

INTERNATIONAL AIR TRANSPORT ASSOCIATION. 2013. *IATA 2013 Report on Alternative Fuels*. Montreal.

INTERNATIONAL AVIATION TRANSPORT ASSOCIATION. 2014. *IATA Fuel Price Monitor*. [online]. [Accessed 16 Oct 2014]. Available from World Wide Web: < HYPERLINK "http://www.iata.org/publications/economics/fuel-monitor/Pages/price-analysis.aspx" http://www.iata.org/publications/economics/fuel-monitor/Pages/price-analysis.aspx >

JOINT RESEARCH CENTRE. 2014. *Well-To-Wheels Analysis of Future Automotive Fuels and Powertrains in the European Context*. Brussels.

KÖHLER, Jonathan, Rainer WALZ, Frank MARSCHEDER-WEIDEMAN, and Benjamin THEDIECK. 2013. *Lead Markets in Aviation Biofuels*. Karlsruhe.

KARLSCH, Rainer and Raymond G. STOKES. 2003. *Faktor Öl. Die Mineralölwirtschaft in Deutschland 1859–1974*. München: C.H. Beck.

KUMP, Lee R., James F. KASTING, and Robert G. CRANE. 2009. *The Earth System*. Pennsylvania: Prentice Hall.

LEE, D.S., G. PITARI, V. GREWE et al. 2010. Transport impacts on atmosphere and climate: Aviation. *Atmospheric Environment*. **44**(1), pp.4678-4734.

LUFTHANSA. 2013. *Lufthansa Annual Report 2013*. Frankfurt am Main.

MANTZOS, L. and P. CAPROS. 2006. *European energy and transport - Scenarios on energy efficiency and renewable*. Brussels.

OLAH, George A., Alain GOEPPERT, and Surya G.K. PRAKASH. 2009. *Beyond Oil and Gas: The Methanol Economy*. Weinheim: WILEY-VCH Verlag GmbH & Co. KGaA.

ORGANIZATION OF THE PETROLEUM EXPORTING COUNTRIES. 2013. *2013 World Oil Outlook*. Vienna.

UNITED NATIONS. 1998. *KYOTO PROTOCOL TO THE UNITED NATIONS FRAMEWORK CONVENTION ON CLIMATE CHANGE.*